Counting Money Workbook

Dr. Michael Stachiw

Counting Money Workbook

Printed in the United States of America

Published by SM&DS through Createspace Independent Publishing Platform

ISBN-13: 978-1523837595

ISBN-10: 1523837594

Two Coin Money Addition - 1

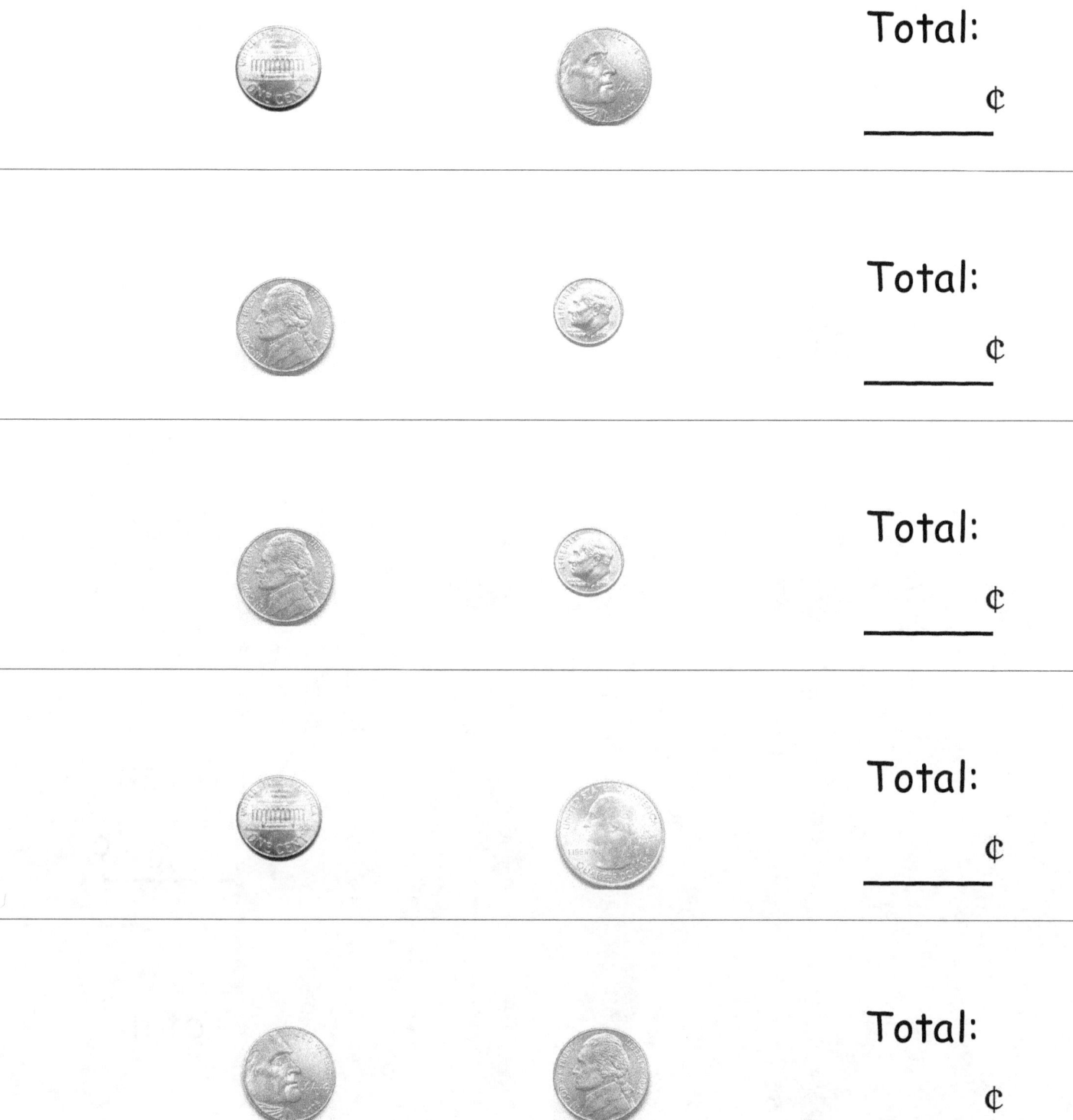

Total:

_____ ¢

Total:

_____ ¢

Total:

_____ ¢

Total:

_____ ¢

Total:

_____ ¢

Two Coin Money Addition - 2

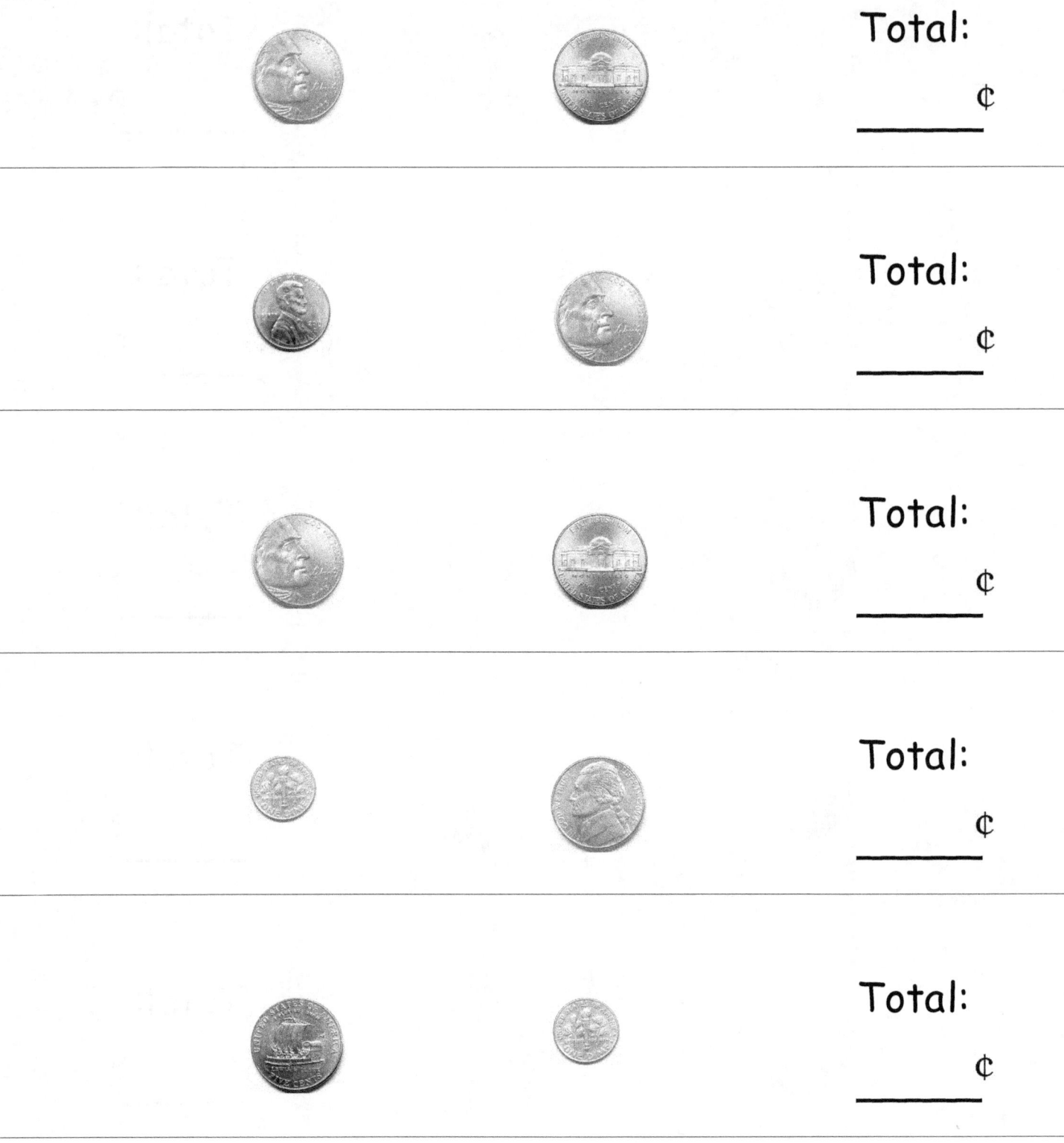

Total:

_____ ¢

Total:

_____ ¢

Total:

_____ ¢

Total:

_____ ¢

Total:

_____ ¢

Two Coin Money Addition - 3

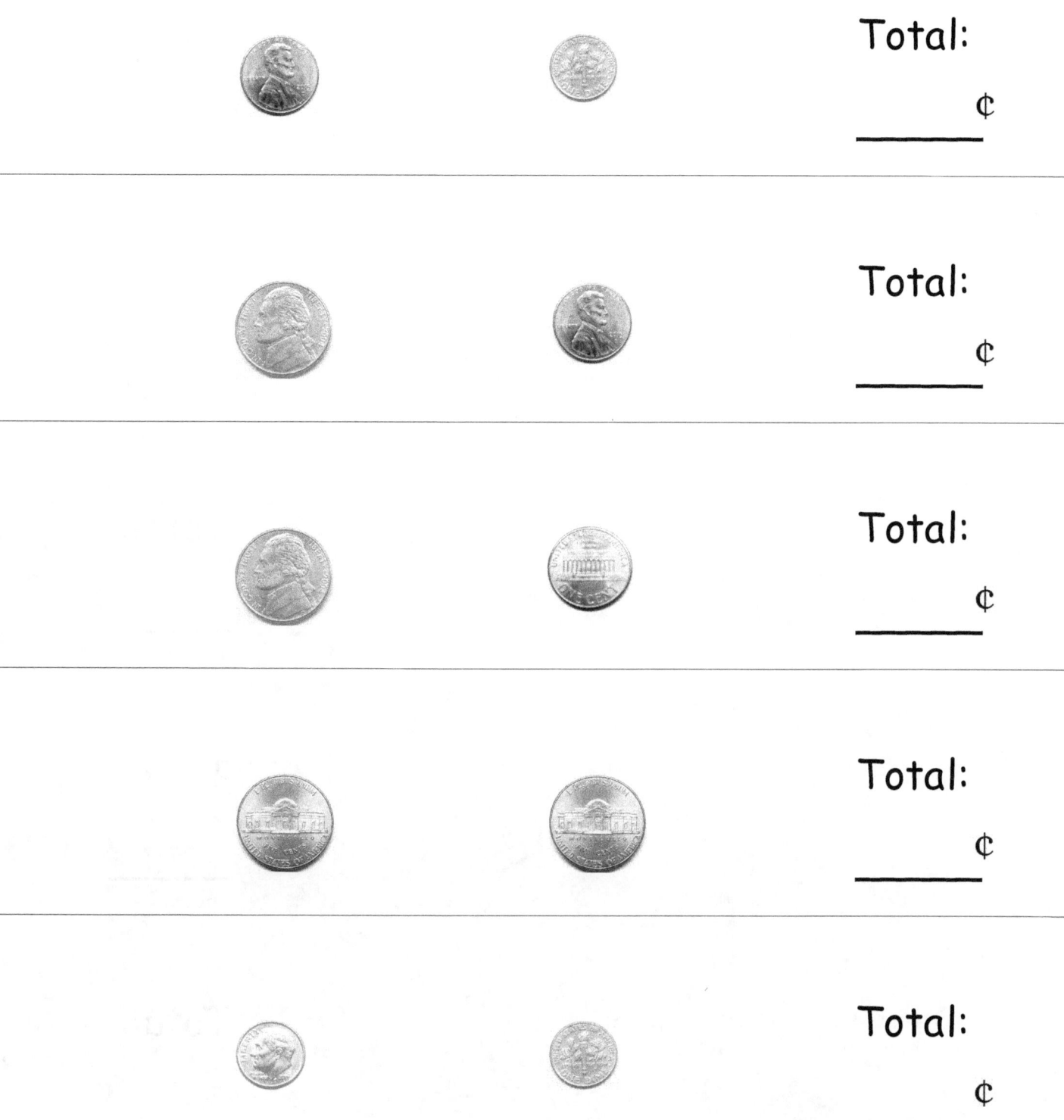

Total:

_____ ¢

Total:

_____ ¢

Total:

_____ ¢

Total:

_____ ¢

Total:

_____ ¢

Two Coin Money Addition - 4

Total:

_____ ¢

Total:

_____ ¢

Total:

_____ ¢

Total:

_____ ¢

Total:

_____ ¢

Two Coin Money Addition - 5

Total:

_____ ¢

Total:

_____ ¢

Total:

_____ ¢

Total:

_____ ¢

Total:

_____ ¢

Two Coin Money Addition - 6

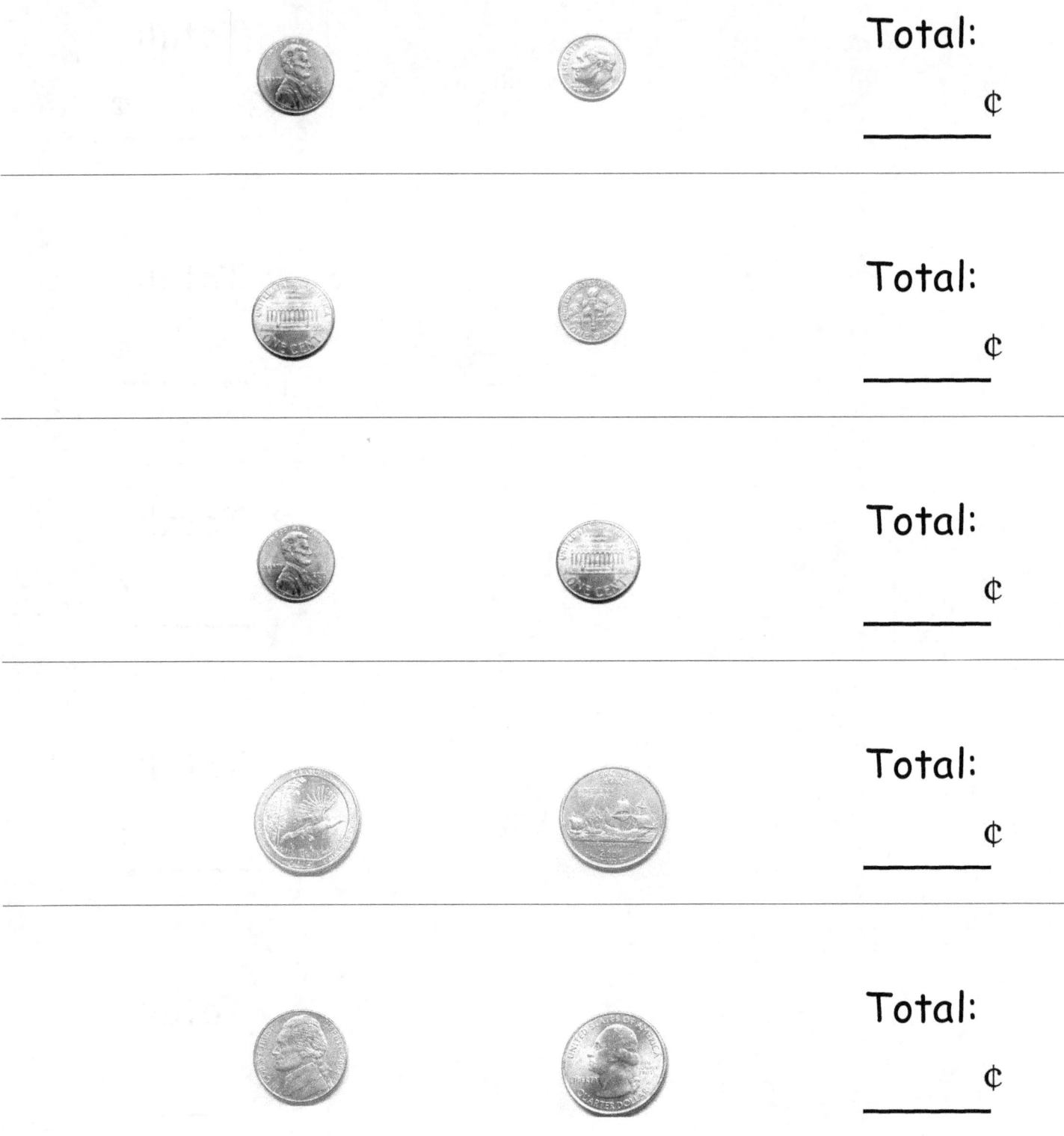

Total:

_____ ¢

Total:

_____ ¢

Total:

_____ ¢

Total:

_____ ¢

Total:

_____ ¢

Two Coin Money Addition - 7

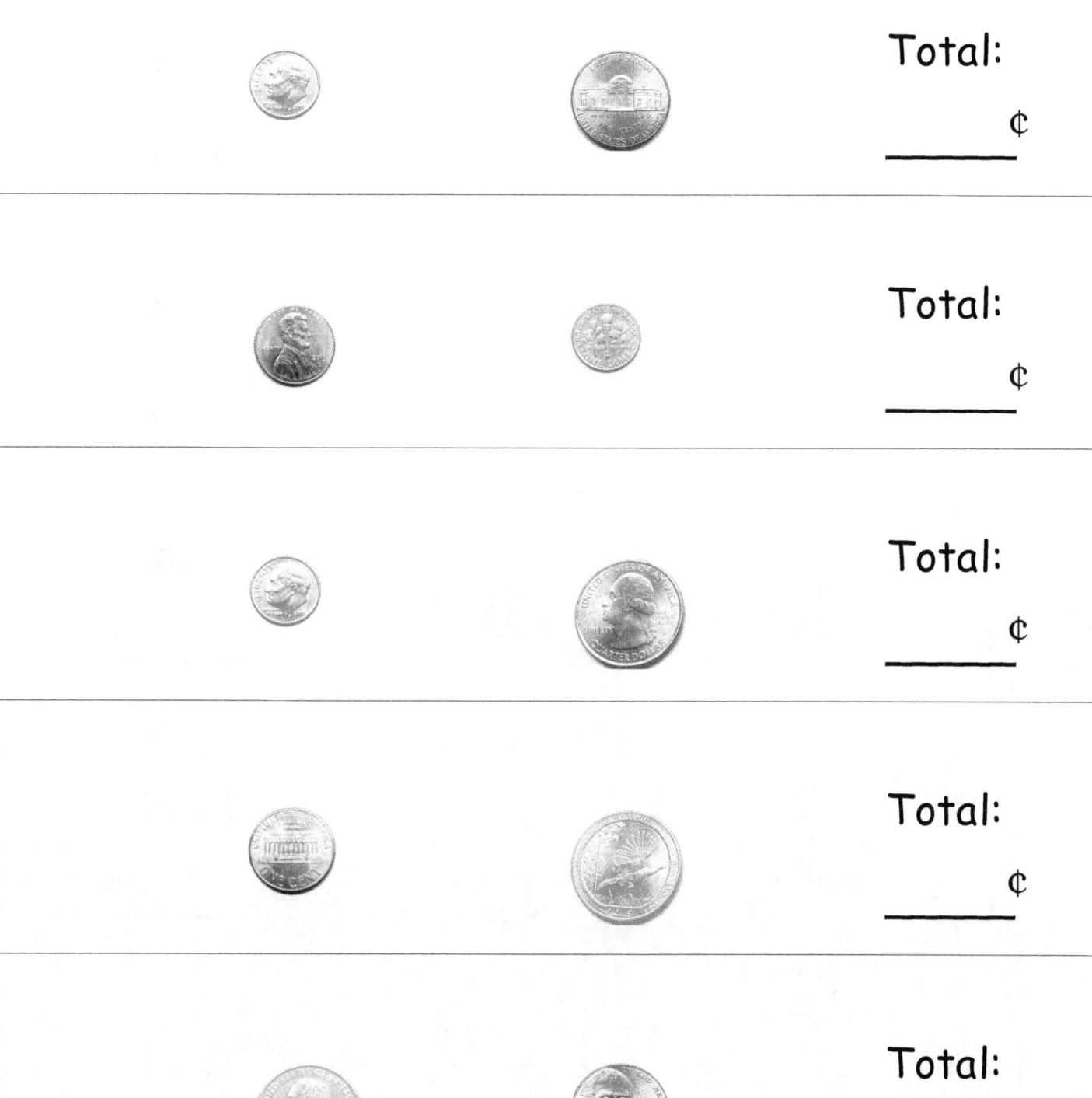

Total:

_____ ¢

Total:

_____ ¢

Total:

_____ ¢

Total:

_____ ¢

Total:

_____ ¢

Two Coin Money Addition - 8

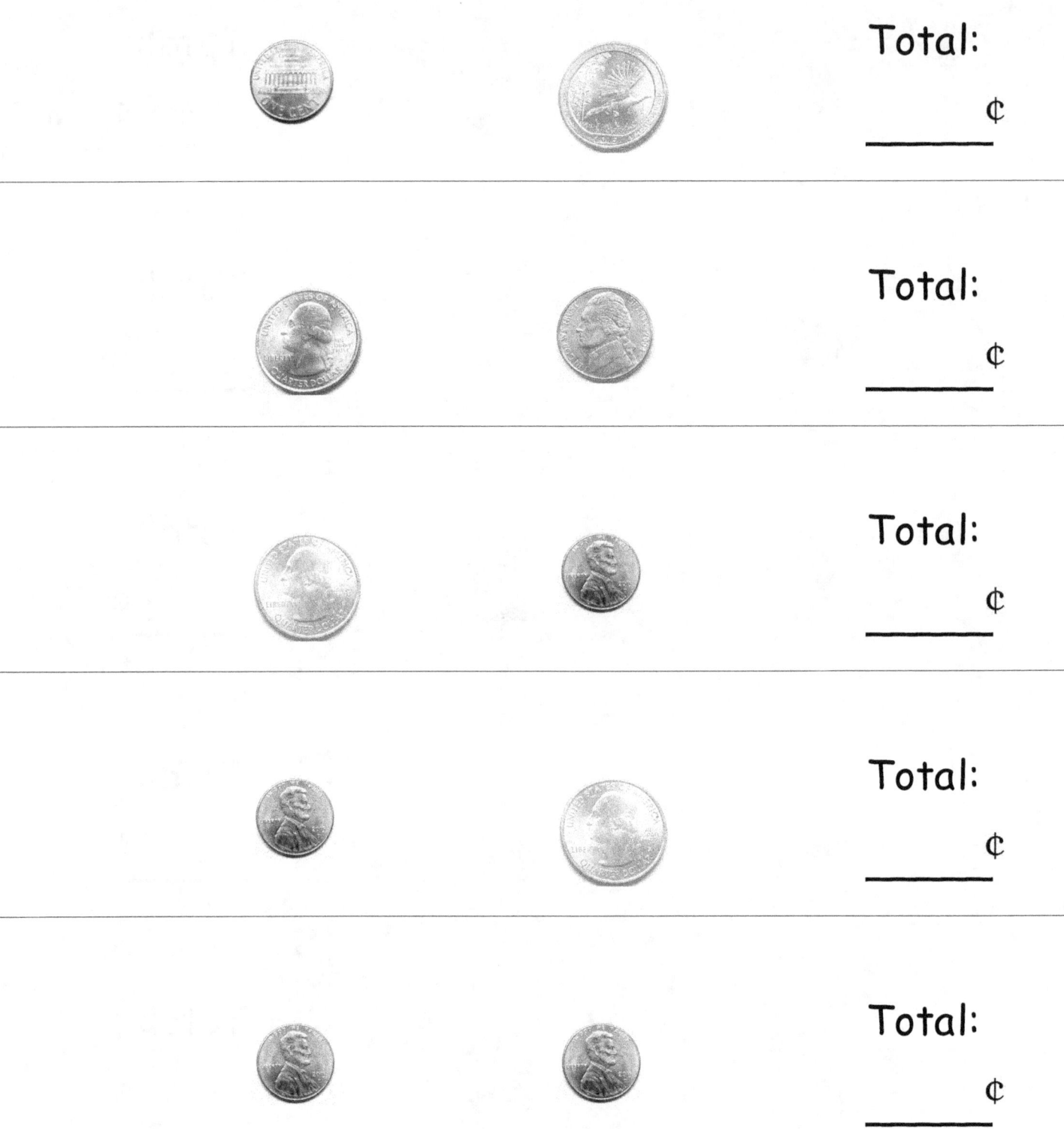

Total:

_____ ¢

Total:

_____ ¢

Total:

_____ ¢

Total:

_____ ¢

Total:

_____ ¢

Two Coin Money Addition - 9

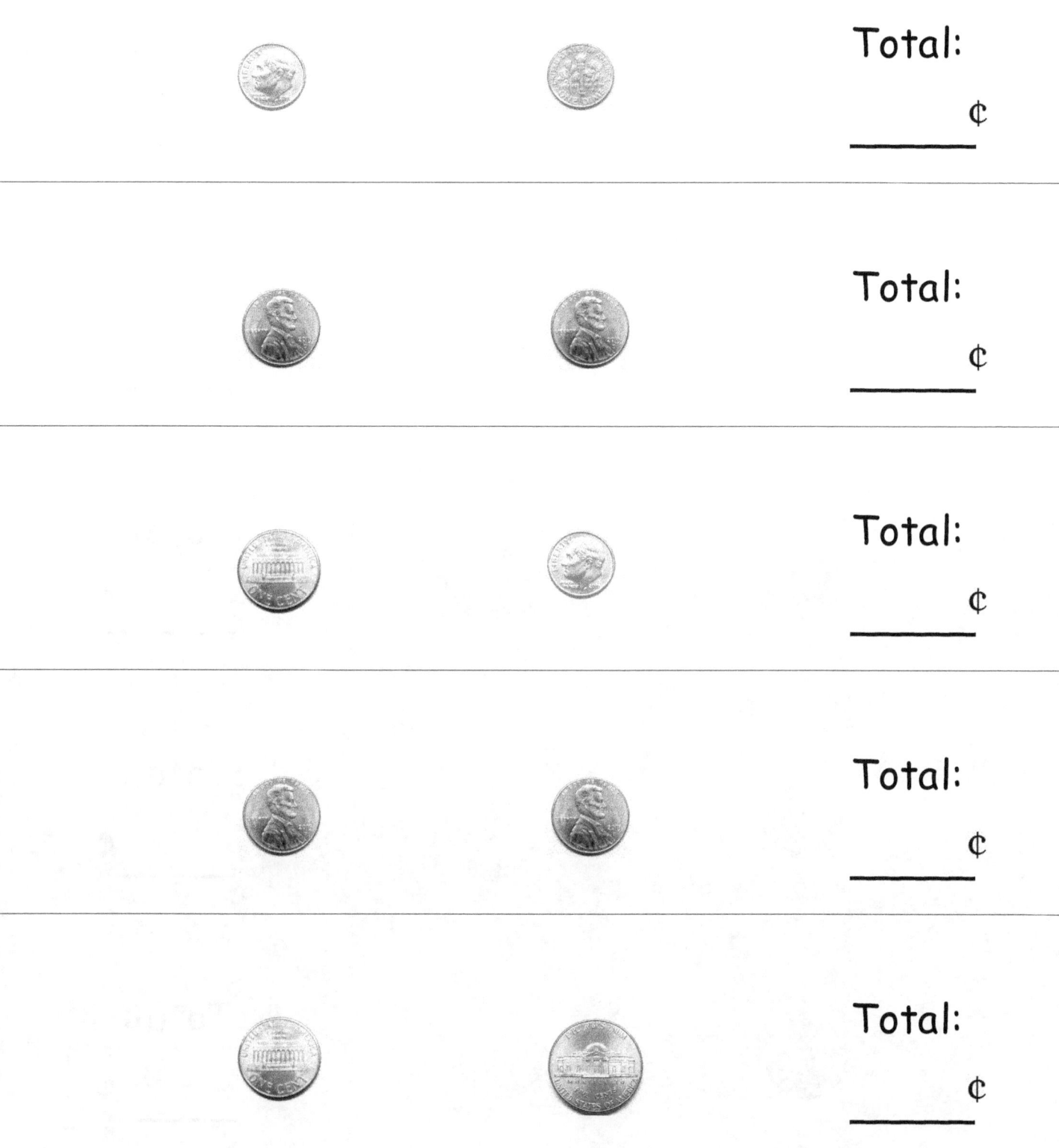

Total:

_____ ¢

Total:

_____ ¢

Total:

_____ ¢

Total:

_____ ¢

Total:

_____ ¢

Two Coin Money Addition - 10

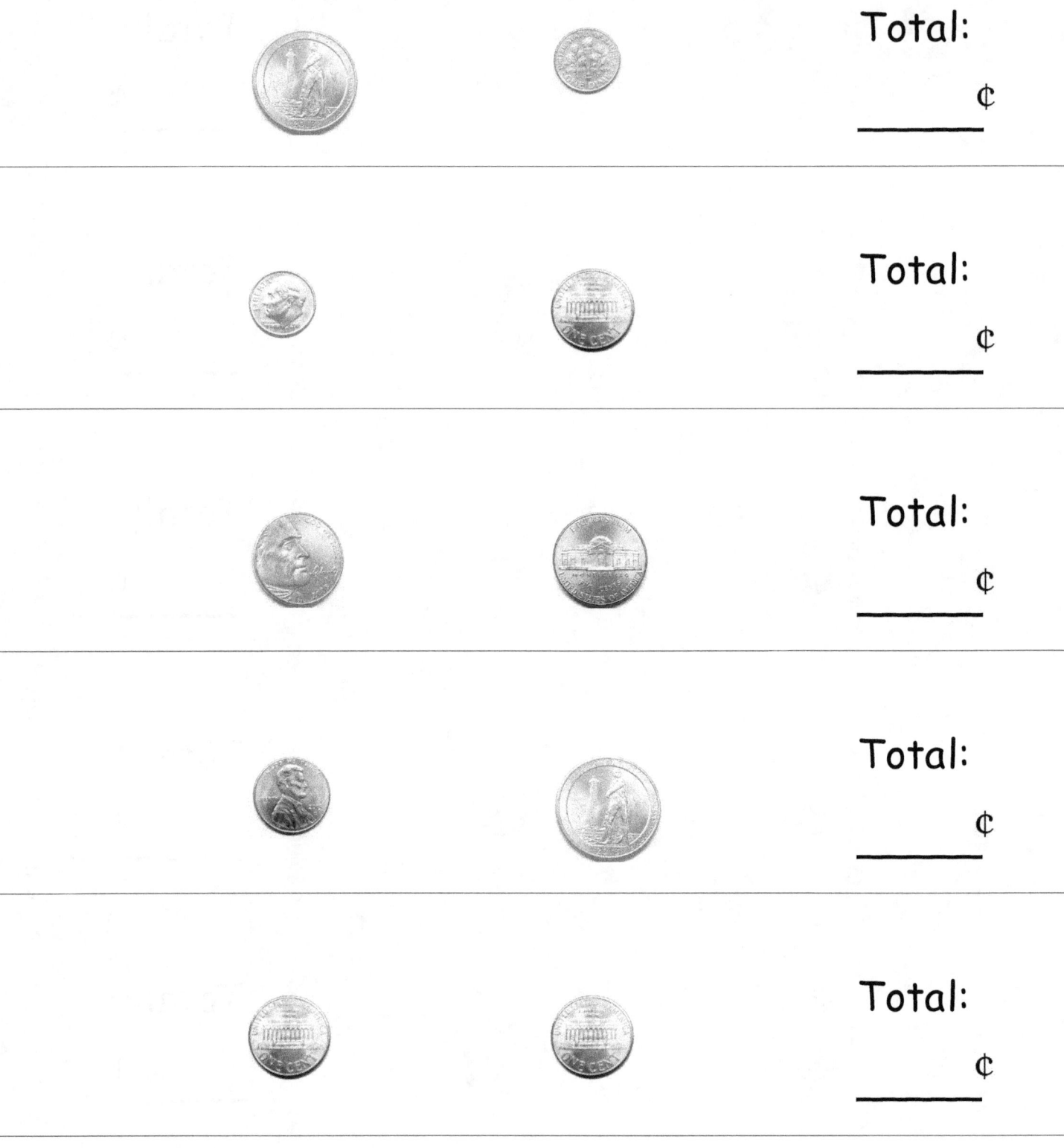

Total:

_____ ¢

Total:

_____ ¢

Total:

_____ ¢

Total:

_____ ¢

Total:

_____ ¢

Three Coin Money Addition - 1

Total:

_____ ¢

Total:

_____ ¢

Total:

_____ ¢

Total:

_____ ¢

Total:

_____ ¢

Three Coin Money Addition - 2

   Total:

_____ ¢

   Total:

_____ ¢

   Total:

_____ ¢

   Total:

_____ ¢

   Total:

_____ ¢

Three Coin Money Addition - 3

Total:

_____ ¢

Total:

_____ ¢

Total:

_____ ¢

Total:

_____ ¢

Total:

_____ ¢

Three Coin Money Addition - 4

   **Total:**

_____ ¢

   **Total:**

_____ ¢

   **Total:**

_____ ¢

   **Total:**

_____ ¢

   **Total:**

_____ ¢

Three Coin Money Addition - 5

Total:

_____ ¢

Total:

_____ ¢

Total:

_____ ¢

Total:

_____ ¢

Total:

_____ ¢

Three Coin Money Addition - 6

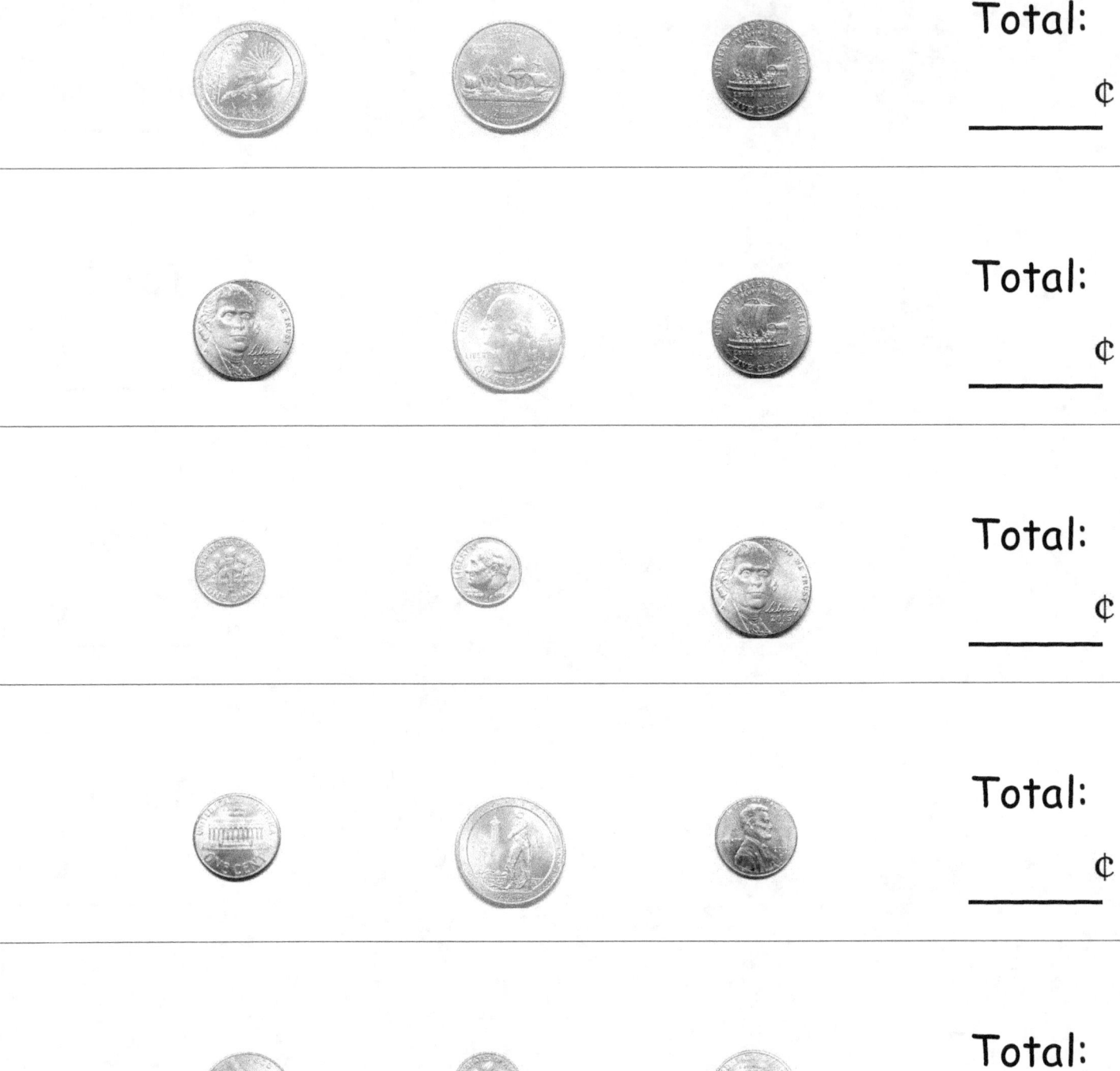

Total: _____ ¢

Total: _____ ¢

Total: _____ ¢

Total: _____ ¢

Total: _____ ¢

Three Coin Money Addition - 7

Total:

_____ ¢

Total:

_____ ¢

Total:

_____ ¢

Total:

_____ ¢

Total:

_____ ¢

Three Coin Money Addition - 8

Total: _____ ¢

Total: _____ ¢

Total: _____ ¢

Total: _____ ¢

Total: _____ ¢

Three Coin Money Addition - 9

   Total:

_____ ¢

   Total:

_____ ¢

   Total:

_____ ¢

   Total:

_____ ¢

   Total:

_____ ¢

Three Coin Money Addition - 10

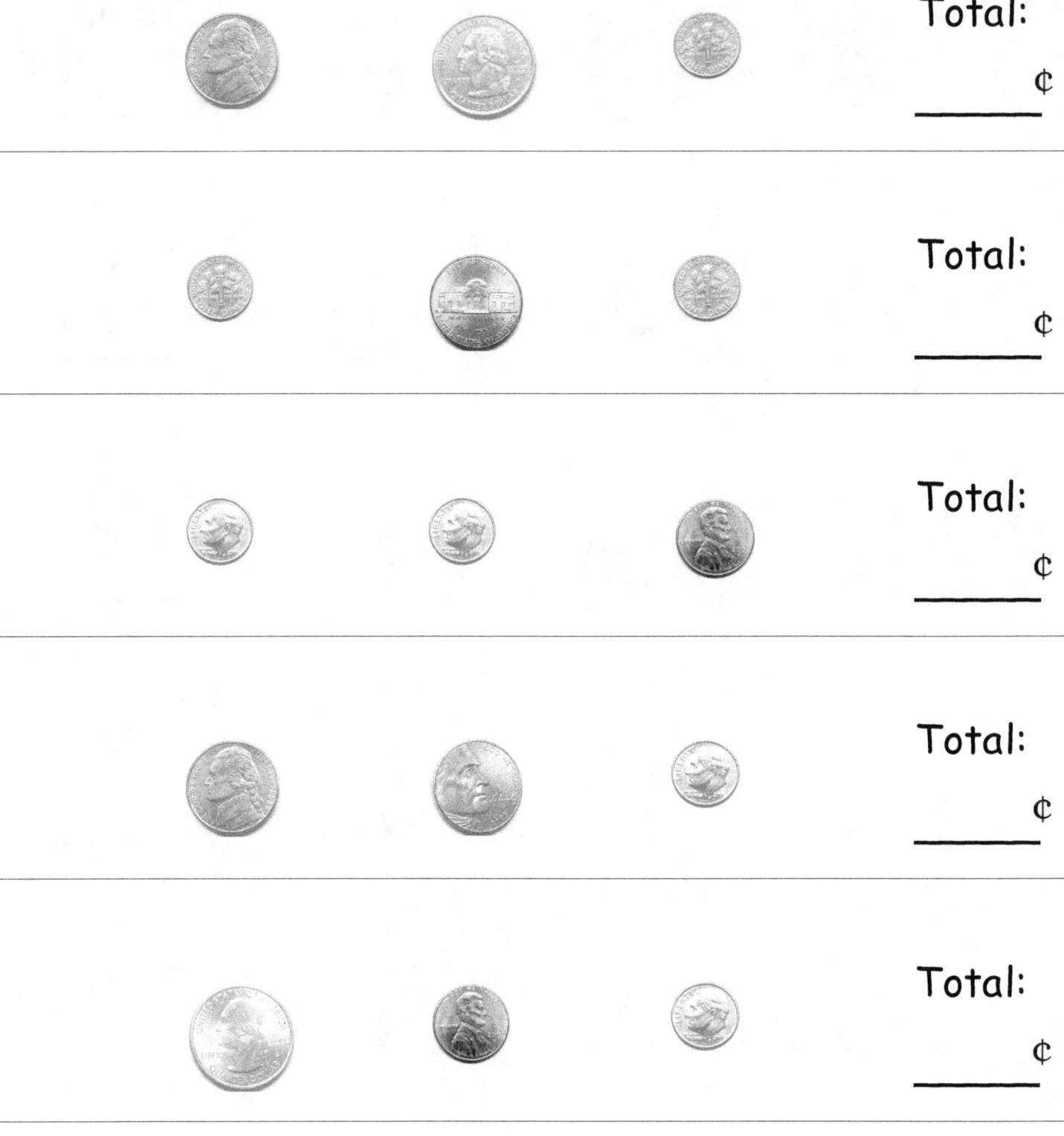

Total:
_____ ¢

Total:
_____ ¢

Total:
_____ ¢

Total:
_____ ¢

Total:
_____ ¢

Four Coin Money Addition - 1

    Total:

_____ ¢

    Total:

_____ ¢

    Total:

_____ ¢

    Total:

_____ ¢

    Total:

_____ ¢

Four Coin Money Addition - 2

    Total:

_____ ¢

    Total:

_____ ¢

    Total:

_____ ¢

    Total:

_____ ¢

    Total:

_____ ¢

Four Coin Money Addition - 3

    Total:

_____ ¢

    Total:

_____ ¢

    Total:

_____ ¢

    Total:

_____ ¢

    Total:

_____ ¢

Four Coin Money Addition - 4

    Total:

_____ ¢

    Total:

_____ ¢

    Total:

_____ ¢

    Total:

_____ ¢

    Total:

_____ ¢

Four Coin Money Addition - 5

Total:

_____ ¢

Total:

_____ ¢

Total:

_____ ¢

Total:

_____ ¢

Total:

_____ ¢

Four Coin Money Addition - 6

      Total:

_____ ¢

      Total:

_____ ¢

      Total:

_____ ¢

      Total:

_____ ¢

      Total:

_____ ¢

Four Coin Money Addition - 7

      Total:

_____ ¢

      Total:

_____ ¢

      Total:

_____ ¢

      Total:

_____ ¢

      Total:

_____ ¢

Four Coin Money Addition - 8

    Total:

_____ ¢

    Total:

_____ ¢

    Total:

_____ ¢

    Total:

_____ ¢

    Total:

_____ ¢

Four Coin Money Addition - 9

    Total:

_____ ¢

    Total:

_____ ¢

    Total:

_____ ¢

    Total:

_____ ¢

    Total:

_____ ¢

Four Coin Money Addition - 10

    Total:

_____ ¢

    Total:

_____ ¢

    Total:

_____ ¢

    Total:

_____ ¢

    Total:

_____ ¢

Five Coin Money Addition - 1

Total:

_____ ¢

Total:

_____ ¢

Total:

_____ ¢

Total:

_____ ¢

Total:

_____ ¢

Five Coin Money Addition - 2

Total:

_____ ¢

Total:

_____ ¢

Total:

_____ ¢

Total:

_____ ¢

Total:

_____ ¢

Five Coin Money Addition - 3

       Total:

_____ ¢

       Total:

_____ ¢

       Total:

_____ ¢

       Total:

_____ ¢

       Total:

_____ ¢

Five Coin Money Addition - 4

Total:

_____ ¢

Total:

_____ ¢

Total:

_____ ¢

Total:

_____ ¢

Total:

_____ ¢

Five Coin Money Addition - 5

Total:

_____ ¢

Total:

_____ ¢

Total:

_____ ¢

Total:

_____ ¢

Total:

_____ ¢

Five Coin Money Addition - 6

       Total:

_____ ¢

       Total:

_____ ¢

       Total:

_____ ¢

       Total:

_____ ¢

       Total:

_____ ¢

Five Coin Money Addition - 7

       Total:

_____ ¢

       Total:

_____ ¢

       Total:

_____ ¢

       Total:

_____ ¢

       Total:

_____ ¢

Five Coin Money Addition - 8

Total:

_____ ¢

Total:

_____ ¢

Total:

_____ ¢

Total:

_____ ¢

Total:

_____ ¢

Five Coin Money Addition - 9

     Total:
_____ ¢

     Total:
_____ ¢

     Total:
_____ ¢

     Total:
_____ ¢

     Total:
_____ ¢

Five Coin Money Addition - 10

     Total:

_____ ¢

     Total:

_____ ¢

     Total:

_____ ¢

     Total:

_____ ¢

     Total:

_____ ¢

Two Coin Money Addition - 1

Total:

6 ¢

Total:

15 ¢

Total:

15 ¢

Total:

26 ¢

Total:

10 ¢

Two Coin Money Addition - 2

Total:

__10__ ¢

Total:

__6__ ¢

Total:

__10__ ¢

Total:

__15__ ¢

Total:

__15__ ¢

Two Coin Money Addition - 3

  Total: ___11___ ¢

  Total: ___6___ ¢

  Total: ___6___ ¢

  Total: ___10___ ¢

  Total: ___20___ ¢

Two Coin Money Addition - 4

　　　　　　Total:

__15__ ¢

　　　　　　Total:

__30__ ¢

　　　　　　Total:

__35__ ¢

　　　　　　Total:

__30__ ¢

　　　　　　Total:

__15__ ¢

Two Coin Money Addition - 5

Total:

__10__ ¢

Total:

__35__ ¢

Total:

__11__ ¢

Total:

__11__ ¢

Total:

__26__ ¢

Two Coin Money Addition - 6

Total:

__11__ ¢

Total:

__11__ ¢

Total:

__2__ ¢

Total:

__50__ ¢

Total:

__30__ ¢

Two Coin Money Addition - 7

  Total:

_15_ ¢

  Total:

_11_ ¢

  Total:

_35_ ¢

  Total:

_26_ ¢

  Total:

_30_ ¢

Two Coin Money Addition - 8

  Total:

    <u> 26 </u> ¢

  Total:

    <u> 30 </u> ¢

  Total:

    <u> 26 </u> ¢

  Total:

    <u> 26 </u> ¢

  Total:

    <u> 2 </u> ¢

Two Coin Money Addition - 9

      Total:

                               20 ¢

      Total:

                               2 ¢

      Total:

                               11 ¢

      Total:

                               2 ¢

      Total:

                               6 ¢

Two Coin Money Addition - 10

  Total:

_35_ ¢

  Total:

_11_ ¢

  Total:

_10_ ¢

  Total:

_26_ ¢

  Total:

_2_ ¢

Three Coin Money Addition - 1

Total:

16 ¢

Total:

51 ¢

Total:

15 ¢

Total:

35 ¢

Total:

40 ¢

Three Coin Money Addition - 2

Total:

51 ¢

Total:

51 ¢

Total:

36 ¢

Total:

21 ¢

Total:

11 ¢

Three Coin Money Addition - 3

   Total:

  20 ¢

   Total:

  45 ¢

   Total:

  51 ¢

   Total:

  21 ¢

   Total:

  31 ¢

Three Coin Money Addition - 4

   Total: __51__ ¢

   Total: __36__ ¢

   Total: __55__ ¢

   Total: __12__ ¢

   Total: __31__ ¢

Three Coin Money Addition - 5

     Total:

_55_ ¢

     Total:

_40_ ¢

     Total:

_16_ ¢

     Total:

_20_ ¢

     Total:

_31_ ¢

Three Coin Money Addition - 6

   Total: 55 ¢

   Total: 35 ¢

   Total: 25 ¢

   Total: 27 ¢

   Total: 16 ¢

Three Coin Money Addition - 7

Total:

_16_ ¢

Total:

_31_ ¢

Total:

_20_ ¢

Total:

_7_ ¢

Total:

_36_ ¢

Three Coin Money Addition - 8

Total:

_45_ ¢

Total:

_11_ ¢

Total:

_55_ ¢

Total:

_3_ ¢

Total:

_36_ ¢

Three Coin Money Addition - 9

Total:

__31__ ¢

Total:

__27__ ¢

Total:

__16__ ¢

Total:

__20__ ¢

Total:

__25__ ¢

Three Coin Money Addition - 10

Total:

___40___ ¢

Total:

___25___ ¢

Total:

___21___ ¢

Total:

___20___ ¢

Total:

___36___ ¢

Four Coin Money Addition - 1

      Total:

_____61_____ ¢

      Total:

_____45_____ ¢

      Total:

_____41_____ ¢

      Total:

_____17_____ ¢

      Total:

_____32_____ ¢

Four Coin Money Addition - 2

    Total:

_____61_____ ¢

    Total:

_____45_____ ¢

    Total:

_____16_____ ¢

    Total:

_____50_____ ¢

    Total:

_____41_____ ¢

Four Coin Money Addition - 3

      Total:

_36_ ¢

      Total:

_100_ ¢

      Total:

_61_ ¢

      Total:

_26_ ¢

      Total:

_50_ ¢

Four Coin Money Addition - 4

Total:

46 ¢

Total:

50 ¢

Total:

50 ¢

Total:

26 ¢

Total:

70 ¢

Four Coin Money Addition - 5

Total:

__17__ ¢

Total:

__26__ ¢

Total:

__36__ ¢

Total:

__21__ ¢

Total:

__46__ ¢

Four Coin Money Addition - 6

    Total:

_16_ ¢

    Total:

_40_ ¢

    Total:

_41_ ¢

    Total:

_21_ ¢

    Total:

_60_ ¢

Four Coin Money Addition - 7

    Total: __56__ ¢

    Total: __32__ ¢

    Total: __46__ ¢

    Total: __45__ ¢

    Total: __56__ ¢

Four Coin Money Addition - 8

    Total: 30 ¢

    Total: 50 ¢

    Total: 56 ¢

    Total: 22 ¢

    Total: 55 ¢

Four Coin Money Addition - 9

Total:

26 ¢

Total:

56 ¢

Total:

35 ¢

Total:

70 ¢

Total:

17 ¢

Four Coin Money Addition - 10

Total:

___21___ ¢

Total:

___36___ ¢

Total:

___61___ ¢

Total:

___30___ ¢

Total:

___40___ ¢

Five Coin Money Addition - 1

Total:

46 ¢

Total:

57 ¢

Total:

41 ¢

Total:

41 ¢

Total:

38 ¢

Five Coin Money Addition - 2

    　Total:

62 ¢

    　Total:

31 ¢

    　Total:

57 ¢

    　Total:

36 ¢

    　Total:

42 ¢

Five Coin Money Addition - 3

Total:

__23__ ¢

Total:

__86__ ¢

Total:

__70__ ¢

Total:

__46__ ¢

Total:

__22__ ¢

Five Coin Money Addition - 4

Total:

__75__ ¢

Total:

__66__ ¢

Total:

__70__ ¢

Total:

__85__ ¢

Total:

__105__ ¢

Five Coin Money Addition - 5

Total:

__56__ ¢

Total:

__27__ ¢

Total:

__81__ ¢

Total:

__18__ ¢

Total:

__70__ ¢

Five Coin Money Addition - 6

Total:

35 ¢

Total:

66 ¢

Total:

18 ¢

Total:

37 ¢

Total:

66 ¢

Five Coin Money Addition - 7

Total:

41 ¢

Total:

40 ¢

Total:

62 ¢

Total:

26 ¢

Total:

46 ¢

Five Coin Money Addition - 8

Total:

66 ¢

Total:

60 ¢

Total:

86 ¢

Total:

51 ¢

Total:

70 ¢

Five Coin Money Addition - 9

Total:

46 ¢

Total:

37 ¢

Total:

57 ¢

Total:

75 ¢

Total:

57 ¢

Five Coin Money Addition - 10

Total:

56 ¢

Total:

47 ¢

Total:

70 ¢

Total:

86 ¢

Total:

56 ¢

About the Book

Count money using any amount of coins. This is an educational workbook for kids to practice counting money.

About the Author

Dr. Stachiw has obtained his Ph.D. in food science from Michigan State University. His formal training is in animal science, agricultural business, and food science. He also majored in statistics and computer mathematical modeling. He is intrigued by numbers and their daily use. You can contact him at Dr.Mike@FeedDealer.com